Name:

Company:

Contact Number:

Contact Number:

Email:

Emergency Details:

Name:

Company:

Name:

Company:

Important:

Date:		**Day:**	Mon	Tue	Wed	Thu	Fri	Sat	Sun

Foreman:

Contract No:

Hours Lost Due To Bad Weather

Visitors

Weather Conditions

AM	PM

Schedule

Completion Date:

Days Ahead of Schedute:

Days Behind Schedute:

Problems / Delays

Safely Issues:

Accidents/Incidents:

Summary Of Work Performed Today

Signature:	Name:

Equipment On Site	No Units	Working	
		Yes	No

Employee /Contractor	Trade	Contracted Hours	Overtime

Materials Delivered	From & Rate	Equipment Rented	No Units

Notes:

| Date: | **Day:** | Mon | Tue | Wed | Thu | Fri | Sat | Sun |

Foreman:

Contract No:

Hours Lost Due To Bad Weather

Visitors

Weather Conditions

AM	PM

Schedule

Completion Date:

Days Ahead of Schedute:

Days Behind Schedute:

Problems / Delays

Safely Issues:

Accidents/Incidents:

Summary Of Work Performed Today

Signature:

Name:

Equipment On Site	No Units	Working	
		Yes	No

Employee /Contractor	Trade	Contracted Hours	Overtime

Materials Delivered	From & Rate	Equipment Rented	No Units

Notes:

Date:		**Day:**	Mon	Tue	Wed	Thu	Fri	Sat	Sun
Foreman:									
Contract No:									

Hours Lost Due To Bad Weather

Visitors

Weather Conditions

AM	PM

Schedule

Completion Date:

Days Ahead of Schedute:

Days Behind Schedute:

Problems / Delays

Safely Issues:

Accidents/Incidents:

Summary Of Work Performed Today

Signature:	Name:

Equipment On Site	No Units	Working	
		Yes	No

Employee /Contractor	Trade	Contracted Hours	Overtime

Materials Delivered	From & Rate	Equipment Rented	No Units

Notes:

Date:		**Day:**	Mon Tue Wed Thu Fri Sat Sun

Foreman:

Contract No:

Hours Lost Due To Bad Weather

Visitors

Weather Conditions

AM	PM

Schedule

Completion Date:

Days Ahead of Schedute:

Days Behind Schedute:

Problems / Delays

Safely Issues:

Accidents/Incidents:

Summary Of Work Performed Today

Signature:

Name:

Equipment On Site	No Units	Working	
		Yes	No

Employee /Contractor	Trade	Contracted Hours	Overtime

Materials Delivered	From & Rate	Equipment Rented	No Units

Notes:

Date:	**Day:**	Mon	Tue	Wed	Thu	Fri	Sat	Sun
Foreman:								
Contract No:								

Hours Lost Due To Bad Weather

Visitors

Weather Conditions

AM	PM

Schedule

Completion Date:

Days Ahead of Schedute:

Days Behind Schedute:

Problems / Delays

Safely Issues:

Accidents/Incidents:

Summary Of Work Performed Today

Signature:

Name:

Equipment On Site	No Units	Working	
		Yes	No

Employee /Contractor	Trade	Contracted Hours	Overtime

Materials Delivered	From & Rate	Equipment Rented	No Units

Notes:

Date:		**Day:**	Mon	Tue	Wed	Thu	Fri	Sat	Sun
Foreman:									
Contract No:									

Hours Lost Due To Bad Weather

Visitors

Weather Conditions

AM	PM

Schedule

Completion Date:

Days Ahead of Schedute:

Days Behind Schedute:

Problems / Delays

Safely Issues:

Accidents/Incidents:

Summary Of Work Performed Today

Signature:	Name:

Equipment On Site	No Units	Working	
		Yes	No

Employee /Contractor	Trade	Contracted Hours	Overtime

Materials Delivered	From & Rate	Equipment Rented	No Units

Notes:

Date:	**Day:**	Mon Tue Wed Thu Fri Sat Sun
Foreman:		
Contract No:		

Hours Lost Due To Bad Weather

Visitors

Weather Conditions

AM	PM

Schedule

Completion Date:

Days Ahead of Schedute:

Days Behind Schedute:

Problems / Delays

Safely Issues:

Accidents/Incidents:

Summary Of Work Performed Today

Signature:	Name:

Equipment On Site	No Units	Working	
		Yes	No

Employee /Contractor	Trade	Contracted Hours	Overtime

Materials Delivered	From & Rate	Equipment Rented	No Units

Notes:

Date:		**Day:**	Mon	Tue	Wed	Thu	Fri	Sat	Sun

Foreman:

Contract No:

Hours Lost Due To Bad Weather

Visitors

Weather Conditions

AM	PM

Schedule

Completion Date:

Days Ahead of Schedute:

Days Behind Schedute:

Problems / Delays

Safely Issues:

Accidents/Incidents:

Summary Of Work Performed Today

Signature:

Name:

Equipment On Site	No Units	Working	
		Yes	No

Employee /Contractor	Trade	Contracted Hours	Overtime

Materials Delivered	From & Rate	Equipment Rented	No Units

Notes:

| Date: | **Day:** | Mon | Tue | Wed | Thu | Fri | Sat | Sun |

Foreman:

Contract No:

Hours Lost Due To Bad Weather

Visitors

Weather Conditions

AM	PM

Schedule

Completion Date:

Days Ahead of Schedute:

Days Behind Schedute:

Problems / Delays

Safely Issues:

Accidents/Incidents:

Summary Of Work Performed Today

Signature:

Name:

Equipment On Site	No Units	Working	
		Yes	No

Employee /Contractor	Trade	Contracted Hours	Overtime

Materials Delivered	From & Rate	Equipment Rented	No Units

Notes:

| Date: | | **Day:** | Mon | Tue | Wed | Thu | Fri | Sat | Sun |

Foreman:

Contract No:

Hours Lost Due To Bad Weather

Visitors

Weather Conditions

AM	PM

Schedule

Completion Date:

Days Ahead of Schedute:

Days Behind Schedute:

Problems / Delays

Safely Issues:

Accidents/Incidents:

Summary Of Work Performed Today

Signature:

Name:

Equipment On Site	No Units	Working	
		Yes	No

Employee /Contractor	Trade	Contracted Hours	Overtime

Materials Delivered	From & Rate	Equipment Rented	No Units

Notes:

Date:		**Day:**	Mon	Tue	Wed	Thu	Fri	Sat	Sun

Foreman:

Contract No:

Hours Lost Due To Bad Weather

Visitors

Weather Conditions

AM	PM

Schedule

Completion Date:

Days Ahead of Schedute:

Days Behind Schedule:

Problems / Delays

Safely Issues:

Accidents/Incidents:

Summary Of Work Performed Today

Signature:	Name:

Equipment On Site	No Units	Working	
		Yes	No

Employee /Contractor	Trade	Contracted Hours	Overtime

Materials Delivered	From & Rate	Equipment Rented	No Units

Notes:

Date:		**Day:**	Mon Tue Wed Thu Fri Sat Sun
Foreman:			
Contract No:			

Hours Lost Due To Bad Weather

Visitors

Weather Conditions

AM	PM

Schedule

Completion Date:

Days Ahead of Schedute:

Days Behind Schedute:

Problems / Delays

Safely Issues:

Accidents/Incidents:

Summary Of Work Performed Today

Signature:	Name:

Equipment On Site	No Units	Working	
		Yes	No

Employee /Contractor	Trade	Contracted Hours	Overtime

Materials Delivered	From & Rate	Equipment Rented	No Units

Notes:

| Date: | | **Day:** | Mon | Tue | Wed | Thu | Fri | Sat | Sun |

Foreman:

Contract No:

Hours Lost Due To Bad Weather

Visitors

Weather Conditions

AM	PM

Schedule

Completion Date:

Days Ahead of Schedute:

Days Behind Schedute:

Problems / Delays

Safely Issues:

Accidents/Incidents:

Summary Of Work Performed Today

Signature:

Name:

Equipment On Site	No Units	Working	
		Yes	No

Employee /Contractor	Trade	Contracted Hours	Overtime

Materials Delivered	From & Rate	Equipment Rented	No Units

Notes:

Date:		**Day:**	Mon	Tue	Wed	Thu	Fri	Sat	Sun
Foreman:									
Contract No:									

Hours Lost Due To Bad Weather

Visitors

Weather Conditions

AM	PM

Schedule

Completion Date:

Days Ahead of Schedute:

Days Behind Schedute:

Problems / Delays

Safely Issues:

Accidents/Incidents:

Summary Of Work Performed Today

Signature:

Name:

Equipment On Site	No Units	Working	
		Yes	No

Employee /Contractor	Trade	Contracted Hours	Overtime

Materials Delivered	From & Rate	Equipment Rented	No Units

Notes:

Date:		**Day:**	Mon	Tue	Wed	Thu	Fri	Sat	Sun
Foreman:									
Contract No:									

Hours Lost Due To Bad Weather

Visitors

Weather Conditions

AM	PM

Schedule

Completion Date:

Days Ahead of Schedute:

Days Behind Schedute:

Problems / Delays

Safely Issues:

Accidents/Incidents:

Summary Of Work Performed Today

Signature:

Name:

Equipment On Site	No Units	Working	
		Yes	No

Employee /Contractor	Trade	Contracted Hours	Overtime

Materials Delivered	From & Rate	Equipment Rented	No Units

Notes:

Date:		**Day:**	Mon	Tue	Wed	Thu	Fri	Sat	Sun
Foreman:									
Contract No:									

Hours Lost Due To Bad Weather

Visitors

Weather Conditions

AM	PM

Schedule

Completion Date:

Days Ahead of Schedute:

Days Behind Schedute:

Problems / Delays

Safely Issues:

Accidents/Incidents:

Summary Of Work Performed Today

Signature:

Name:

Equipment On Site	No Units	Working	
		Yes	No

Employee /Contractor	Trade	Contracted Hours	Overtime

Materials Delivered	From & Rate	Equipment Rented	No Units

Notes:

Date:		**Day:**	Mon	Tue	Wed	Thu	Fri	Sat	Sun

Foreman:

Contract No:

Hours Lost Due To Bad Weather

Visitors

Weather Conditions

AM	PM

Schedule

Completion Date:

Days Ahead of Schedute:

Days Behind Schedute:

Problems / Delays

Safely Issues:

Accidents/Incidents:

Summary Of Work Performed Today

Signature:	Name:

Equipment On Site	No Units	Working	
		Yes	No

Employee /Contractor	Trade	Contracted Hours	Overtime

Materials Delivered	From & Rate	Equipment Rented	No Units

Notes:

Date:	**Day:**	Mon	Tue	Wed	Thu	Fri	Sat	Sun

Foreman:

Contract No:

Hours Lost Due To Bad Weather

Visitors

Weather Conditions

AM	PM

Schedule

Completion Date:

Days Ahead of Schedute:

Days Behind Schedute:

Problems / Delays

Safely Issues:

Accidents/Incidents:

Summary Of Work Performed Today

Signature:

Name:

Equipment On Site	No Units	Working	
		Yes	No

Employee /Contractor	Trade	Contracted Hours	Overtime

Materials Delivered	From & Rate	Equipment Rented	No Units

Notes:

| Date: | | **Day:** | Mon | Tue | Wed | Thu | Fri | Sat | Sun |

| Foreman: |
| Contract No: |

Hours Lost Due To Bad Weather

Visitors

Weather Conditions

AM	PM

Schedule

| Completion Date: |
| Days Ahead of Schedute: |
| Days Behind Schedute: |

Problems / Delays

Safely Issues:

Accidents/Incidents:

Summary Of Work Performed Today

| Signature: | Name: |

Equipment On Site	No Units	Working	
		Yes	No

Employee /Contractor	Trade	Contracted Hours	Overtime

Materials Delivered	From & Rate	Equipment Rented	No Units

Notes:

Date:		**Day:**	Mon	Tue	Wed	Thu	Fri	Sat	Sun
Foreman:									
Contract No:									

Hours Lost Due To Bad Weather

Visitors

Weather Conditions

AM	PM

Schedule

Completion Date:

Days Ahead of Schedute:

Days Behind Schedute:

Problems / Delays

Safely Issues:

Accidents/Incidents:

Summary Of Work Performed Today

Signature:

Name:

Equipment On Site	No Units	Working	
		Yes	No

Employee /Contractor	Trade	Contracted Hours	Overtime

Materials Delivered	From & Rate	Equipment Rented	No Units

Notes:

Date:		**Day:**	Mon	Tue	Wed	Thu	Fri	Sat	Sun

Foreman:

Contract No:

Hours Lost Due To Bad Weather

Visitors

Weather Conditions

AM	PM

Schedule

Completion Date:

Days Ahead of Schedute:

Days Behind Schedute:

Problems / Delays

Safely Issues:

Accidents/Incidents:

Summary Of Work Performed Today

Signature:

Name:

Equipment On Site	No Units	Working	
		Yes	No

Employee /Contractor	Trade	Contracted Hours	Overtime

Materials Delivered	From & Rate	Equipment Rented	No Units

Notes:

Date:		**Day:**	Mon Tue Wed Thu Fri Sat Sun

Foreman:	
Contract No:	

Hours Lost Due To Bad Weather

Visitors

Weather Conditions

AM	PM

Schedule

Completion Date:	
Days Ahead of Schedute:	
Days Behind Schedute:	

Problems / Delays

Safely Issues:

Accidents/Incidents:

Summary Of Work Performed Today

Signature:	Name:

Equipment On Site	No Units	Working	
		Yes	No

Employee /Contractor	Trade	Contracted Hours	Overtime

Materials Delivered	From & Rate	Equipment Rented	No Units

Notes:

| Date: | | **Day:** | Mon | Tue | Wed | Thu | Fri | Sat | Sun |

Foreman:

Contract No:

Hours Lost Due To Bad Weather

Visitors

Weather Conditions

AM	PM

Schedule

Completion Date:

Days Ahead of Schedute:

Days Behind Schedule:

Problems / Delays

Safely Issues:

Accidents/Incidents:

Summary Of Work Performed Today

Signature:

Name:

Equipment On Site	No Units	Working	
		Yes	No

Employee /Contractor	Trade	Contracted Hours	Overtime

Materials Delivered	From & Rate	Equipment Rented	No Units

Notes:

Date:	**Day:**	Mon	Tue	Wed	Thu	Fri	Sat	Sun
Foreman:								
Contract No:								

Hours Lost Due To Bad Weather

Visitors

Weather Conditions

AM	PM

Schedule

Completion Date:

Days Ahead of Schedute:

Days Behind Schedute:

Problems / Delays

Safely Issues:

Accidents/Incidents:

Summary Of Work Performed Today

Signature:

Name:

Equipment On Site	No Units	Working	
		Yes	No

Employee /Contractor	Trade	Contracted Hours	Overtime

Materials Delivered	From & Rate	Equipment Rented	No Units

Notes:

Date:		**Day:**	Mon	Tue	Wed	Thu	Fri	Sat	Sun

Foreman:

Contract No:

Hours Lost Due To Bad Weather

Visitors

Weather Conditions

AM	PM

Schedule

Completion Date:

Days Ahead of Schedute:

Days Behind Schedute:

Problems / Delays

Safely Issues:

Accidents/Incidents:

Summary Of Work Performed Today

Signature:	Name:

Equipment On Site	No Units	Working	
		Yes	No

Employee /Contractor	Trade	Contracted Hours	Overtime

Materials Delivered	From & Rate	Equipment Rented	No Units

Notes:

| Date: | | **Day:** | Mon | Tue | Wed | Thu | Fri | Sat | Sun |

Foreman:

Contract No:

Hours Lost Due To Bad Weather

Visitors

Weather Conditions

AM	PM

Schedule

Completion Date:

Days Ahead of Schedute:

Days Behind Schedute:

Problems / Delays

Safely Issues:

Accidents/Incidents:

Summary Of Work Performed Today

Signature:

Name:

Equipment On Site	No Units	Working	
		Yes	No

Employee /Contractor	Trade	Contracted Hours	Overtime

Materials Delivered	From & Rate	Equipment Rented	No Units

Notes:

Date:	**Day:**	Mon	Tue	Wed	Thu	Fri	Sat	Sun

Foreman:

Contract No:

Hours Lost Due To Bad Weather

Visitors

Weather Conditions

AM	PM

Schedule

Completion Date:
Days Ahead of Schedute:
Days Behind Schedute:

Problems / Delays

Safely Issues:

Accidents/Incidents:

Summary Of Work Performed Today

Signature:	Name:

Equipment On Site	No Units	Working	
		Yes	No

Employee /Contractor	Trade	Contracted Hours	Overtime

Materials Delivered	From & Rate	Equipment Rented	No Units

Notes:

Date:		**Day:**	Mon	Tue	Wed	Thu	Fri	Sat	Sun
Foreman:									
Contract No:									

Hours Lost Due To Bad Weather | Visitors

Weather Conditions

AM	PM

Schedule | Problems / Delays

Completion Date:

Days Ahead of Schedute:

Days Behind Schedute:

Safely Issues: | Accidents/Incidents:

Summary Of Work Performed Today

Signature:	Name:

Equipment On Site	No Units	Working	
		Yes	No

Employee /Contractor	Trade	Contracted Hours	Overtime

Materials Delivered	From & Rate	Equipment Rented	No Units
			Yes No

Notes:

Date:		**Day:**	Mon	Tue	Wed	Thu	Fri	Sat	Sun

Foreman:

Contract No:

Hours Lost Due To Bad Weather

Visitors

Weather Conditions

AM	PM

Schedule

Completion Date:

Days Ahead of Schedute:

Days Behind Schedute:

Problems / Delays

Safely Issues:

Accidents/Incidents:

Summary Of Work Performed Today

Signature:

Name:

Equipment On Site	No Units	Working	
		Yes	No

Employee /Contractor	Trade	Contracted Hours	Overtime

Materials Delivered	From & Rate	Equipment Rented	No Units

Notes:

Date:	**Day:**	Mon	Tue	Wed	Thu	Fri	Sat	Sun

Foreman:

Contract No:

Hours Lost Due To Bad Weather

Visitors

Weather Conditions

AM	PM

Schedule

Completion Date:
Days Ahead of Schedute:
Days Behind Schedute:

Problems / Delays

Safely Issues:

Accidents/Incidents:

Summary Of Work Performed Today

Signature:	Name:

Equipment On Site	No Units	Working	
		Yes	No

Employee /Contractor	Trade	Contracted Hours	Overtime

Materials Delivered	From & Rate	Equipment Rented	No Units

Notes:

Date:		**Day:**	Mon	Tue	Wed	Thu	Fri	Sat	Sun
Foreman:									
Contract No:									

Hours Lost Due To Bad Weather

Visitors

Weather Conditions

AM	PM

Schedule

Completion Date:
Days Ahead of Schedute:
Days Behind Schedute:

Problems / Delays

Safely Issues:

Accidents/Incidents:

Summary Of Work Performed Today

Signature:	Name:

Equipment On Site	No Units	Working	
		Yes	No

Employee /Contractor	Trade	Contracted Hours	Overtime

Materials Delivered	From & Rate	Equipment Rented	No Units

Notes:

Date:		**Day:**	Mon	Tue	Wed	Thu	Fri	Sat	Sun

Foreman:

Contract No:

Hours Lost Due To Bad Weather	Visitors

Weather Conditions

AM	PM

Schedule	Problems / Delays
Completion Date:	
Days Ahead of Schedute:	
Days Behind Schedute:	

Safely Issues:	Accidents/Incidents:

Summary Of Work Performed Today

Signature:	Name:

Equipment On Site	No Units	Working	
		Yes	No

Employee /Contractor	Trade	Contracted Hours	Overtime

Materials Delivered	From & Rate	Equipment Rented	No Units

Notes:

Date:		**Day:**	Mon	Tue	Wed	Thu	Fri	Sat	Sun
Foreman:									
Contract No:									

Hours Lost Due To Bad Weather

Visitors

Weather Conditions

AM	PM

Schedule

Completion Date:

Days Ahead of Schedute:

Days Behind Schedute:

Problems / Delays

Safely Issues:

Accidents/Incidents:

Summary Of Work Performed Today

Signature:

Name:

Equipment On Site	No Units	Working	
		Yes	No

Employee /Contractor	Trade	Contracted Hours	Overtime

Materials Delivered	From & Rate	Equipment Rented	No Units

Notes:

Date:		**Day:**	Mon	Tue	Wed	Thu	Fri	Sat	Sun

Foreman:

Contract No:

Hours Lost Due To Bad Weather

Visitors

Weather Conditions

AM	PM

Schedule

Completion Date:

Days Ahead of Schedute:

Days Behind Schedute:

Problems / Delays

Safely Issues:

Accidents/Incidents:

Summary Of Work Performed Today

Signature:

Name:

Equipment On Site	No Units	Working	
		Yes	No

Employee /Contractor	Trade	Contracted Hours	Overtime

Materials Delivered	From & Rate	Equipment Rented	No Units

Notes:

Date:	**Day:**	Mon	Tue	Wed	Thu	Fri	Sat	Sun
Foreman:								
Contract No:								

Hours Lost Due To Bad Weather

Visitors

Weather Conditions

AM	PM

Schedule

Completion Date:

Days Ahead of Schedute:

Days Behind Schedute:

Problems / Delays

Safely Issues:

Accidents/Incidents:

Summary Of Work Performed Today

Signature:	Name:

Equipment On Site	No Units	Working	
		Yes	No

Employee /Contractor	Trade	Contracted Hours	Overtime

Materials Delivered	From & Rate	Equipment Rented	No Units

Notes:

Date:		**Day:**	Mon	Tue	Wed	Thu	Fri	Sat	Sun

Foreman:

Contract No:

Hours Lost Due To Bad Weather

Visitors

Weather Conditions

AM	PM

Schedule

Completion Date:

Days Ahead of Schedute:

Days Behind Schedute:

Problems / Delays

Safely Issues:

Accidents/Incidents:

Summary Of Work Performed Today

Signature:

Name:

Equipment On Site	No Units	Working	
		Yes	No

Employee /Contractor	Trade	Contracted Hours	Overtime

Materials Delivered	From & Rate	Equipment Rented	No Units

Notes:

| Date: | | **Day:** | Mon | Tue | Wed | Thu | Fri | Sat | Sun |

| Foreman: | |

| Contract No: | |

Hours Lost Due To Bad Weather

Visitors

Weather Conditions

AM	PM

Schedule

Completion Date:

Days Ahead of Schedute:

Days Behind Schedute:

Problems / Delays

Safely Issues:

Accidents/Incidents:

Summary Of Work Performed Today

| Signature: | Name: |

Equipment On Site	No Units	Working	
		Yes	No

Employee /Contractor	Trade	Contracted Hours	Overtime

Materials Delivered	From & Rate	Equipment Rented	No Units

Notes:

Date:	Day: Mon Tue Wed Thu Fri Sat Sun
Foreman:	
Contract No:	

Hours Lost Due To Bad Weather

Visitors

Weather Conditions

AM	PM

Schedule

Completion Date:

Days Ahead of Schedute:

Days Behind Schedute:

Problems / Delays

Safely Issues:

Accidents/Incidents:

Summary Of Work Performed Today

Signature:	Name:

Equipment On Site	No Units	Working	
		Yes	No

Employee /Contractor	Trade	Contracted Hours	Overtime

Materials Delivered	From & Rate	Equipment Rented	No Units

Notes:

| Date: | | **Day:** | Mon | Tue | Wed | Thu | Fri | Sat | Sun |

Foreman:

Contract No:

Hours Lost Due To Bad Weather

Visitors

Weather Conditions

AM	PM

Schedule

Completion Date:

Days Ahead of Schedute:

Days Behind Schedute:

Problems / Delays

Safely Issues:

Accidents/Incidents:

Summary Of Work Performed Today

Signature:

Name:

Equipment On Site	No Units	Working	
		Yes	No

Employee /Contractor	Trade	Contracted Hours	Overtime

Materials Delivered	From & Rate	Equipment Rented	No Units

Notes:

Date:		**Day:**	Mon	Tue	Wed	Thu	Fri	Sat	Sun
Foreman:									
Contract No:									

Hours Lost Due To Bad Weather

Visitors

Weather Conditions

AM	PM

Schedule

Completion Date:

Days Ahead of Schedute:

Days Behind Schedute:

Problems / Delays

Safely Issues:

Accidents/Incidents:

Summary Of Work Performed Today

Signature:	Name:

Equipment On Site	No Units	Working	
		Yes	No

Employee /Contractor	Trade	Contracted Hours	Overtime

Materials Delivered	From & Rate	Equipment Rented	No Units

Notes:

Date:	**Day:**	Mon	Tue	Wed	Thu	Fri	Sat	Sun

Foreman:

Contract No:

Hours Lost Due To Bad Weather

Visitors

Weather Conditions

AM	PM

Schedule

Completion Date:

Days Ahead of Schedute:

Days Behind Schedute:

Problems / Delays

Safely Issues:

Accidents/Incidents:

Summary Of Work Performed Today

Signature:	Name:

Equipment On Site	No Units	Working	
		Yes	No

Employee /Contractor	Trade	Contracted Hours	Overtime

Materials Delivered	From & Rate	Equipment Rented	No Units

Notes:

Date:		**Day:**	Mon	Tue	Wed	Thu	Fri	Sat	Sun
Foreman:									
Contract No:									

Hours Lost Due To Bad Weather

Visitors

Weather Conditions

AM	PM

Schedule

Completion Date:

Days Ahead of Schedute:

Days Behind Schedute:

Problems / Delays

Safely Issues:

Accidents/Incidents:

Summary Of Work Performed Today

Signature:

Name:

Equipment On Site	No Units	Working	
		Yes	No

Employee /Contractor	Trade	Contracted Hours	Overtime

Materials Delivered	From & Rate	Equipment Rented	No Units

Notes:

Date:		**Day:**	Mon	Tue	Wed	Thu	Fri	Sat	Sun

Foreman:

Contract No:

Hours Lost Due To Bad Weather	Visitors

Weather Conditions

AM	PM

Schedule	Problems / Delays
Completion Date:	
Days Ahead of Schedute:	
Days Behind Schedule:	

Safely Issues:	Accidents/Incidents:

Summary Of Work Performed Today

Signature:	Name:

Equipment On Site	No Units	Working	
		Yes	No

Employee /Contractor	Trade	Contracted Hours	Overtime

Materials Delivered	From & Rate	Equipment Rented	No Units
			Yes / No

Notes:

Date:	**Day:**	Mon	Tue	Wed	Thu	Fri	Sat	Sun

Foreman:

Contract No:

Hours Lost Due To Bad Weather

Visitors

Weather Conditions

AM	PM

Schedule

Completion Date:

Days Ahead of Schedute:

Days Behind Schedute:

Problems / Delays

Safely Issues:

Accidents/Incidents:

Summary Of Work Performed Today

Signature:

Name:

Equipment On Site	No Units	Working	
		Yes	No

Employee /Contractor	Trade	Contracted Hours	Overtime

Materials Delivered	From & Rate	Equipment Rented	No Units

Notes:

Date:		**Day:**	Mon	Tue	Wed	Thu	Fri	Sat	Sun

Foreman:

Contract No:

Hours Lost Due To Bad Weather	Visitors

Weather Conditions

AM	PM	

Schedule	Problems / Delays
Completion Date:	
Days Ahead of Schedute:	
Days Behind Schedute:	

Safely Issues:	Accidents/Incidents:

Summary Of Work Performed Today

Signature:	Name:

Equipment On Site	No Units	Working	
		Yes	No

Employee /Contractor	Trade	Contracted Hours	Overtime

Materials Delivered	From & Rate	Equipment Rented	No Units

Notes:

Date:		**Day:**	Mon	Tue	Wed	Thu	Fri	Sat	Sun
Foreman:									
Contract No:									

Hours Lost Due To Bad Weather

Visitors

Weather Conditions

AM	PM

Schedule

Completion Date:

Days Ahead of Schedute:

Days Behind Schedute:

Problems / Delays

Safely Issues:

Accidents/Incidents:

Summary Of Work Performed Today

Signature:

Name:

Equipment On Site	No Units	Working	
		Yes	No

Employee /Contractor	Trade	Contracted Hours	Overtime

Materials Delivered	From & Rate	Equipment Rented	No Units

Notes:

Date:		**Day:**	Mon	Tue	Wed	Thu	Fri	Sat	Sun
Foreman:									
Contract No:									

Hours Lost Due To Bad Weather	Visitors

Weather Conditions

AM	PM

Schedule	Problems / Delays
Completion Date:	
Days Ahead of Schedute:	
Days Behind Schedute:	

Safely Issues:	Accidents/Incidents:

Summary Of Work Performed Today

Signature:	Name:

Equipment On Site	No Units	Working	
		Yes	No

Employee /Contractor	Trade	Contracted Hours	Overtime

Materials Delivered	From & Rate	Equipment Rented	No Units

Notes:

Date:		**Day:**	Mon	Tue	Wed	Thu	Fri	Sat	Sun

Foreman:

Contract No:

Hours Lost Due To Bad Weather

Visitors

Weather Conditions

AM	PM

Schedule

Completion Date:

Days Ahead of Schedute:

Days Behind Schedute:

Problems / Delays

Safely Issues:

Accidents/Incidents:

Summary Of Work Performed Today

Signature:

Name:

Equipment On Site	No Units	Working	
		Yes	No

Employee /Contractor	Trade	Contracted Hours	Overtime

Materials Delivered	From & Rate	Equipment Rented	No Units

Notes:

Date:	Day:	Mon	Tue	Wed	Thu	Fri	Sat	Sun

Foreman:

Contract No:

Hours Lost Due To Bad Weather

Visitors

Weather Conditions

AM	PM

Schedule

Completion Date:

Days Ahead of Schedute:

Days Behind Schedute:

Problems / Delays

Safely Issues:

Accidents/Incidents:

Summary Of Work Performed Today

Signature:

Name:

Equipment On Site	No Units	Working Yes	No

Employee /Contractor	Trade	Contracted Hours	Overtime

Materials Delivered	From & Rate	Equipment Rented	No Units

Notes:

| Date: | | **Day:** | Mon | Tue | Wed | Thu | Fri | Sat | Sun |

Foreman:

Contract No:

Hours Lost Due To Bad Weather

Visitors

Weather Conditions

AM	PM

Schedule

Completion Date:

Days Ahead of Schedute:

Days Behind Schedule:

Problems / Delays

Safely Issues:

Accidents/Incidents:

Summary Of Work Performed Today

Signature:

Name:

Equipment On Site	No Units	Working	
		Yes	No

Employee /Contractor	Trade	Contracted Hours	Overtime

Materials Delivered	From & Rate	Equipment Rented	No Units

Notes:

Date:		**Day:**	Mon	Tue	Wed	Thu	Fri	Sat	Sun
Foreman:									
Contract No:									

Hours Lost Due To Bad Weather

Visitors

Weather Conditions

AM	PM

Schedule

Completion Date:

Days Ahead of Schedute:

Days Behind Schedute:

Problems / Delays

Safely Issues:

Accidents/Incidents:

Summary Of Work Performed Today

Signature:

Name:

Equipment On Site	No Units	Working	
		Yes	No

Employee /Contractor	Trade	Contracted Hours	Overtime

Materials Delivered	From & Rate	Equipment Rented	No Units

Notes:

Date:		**Day:**	Mon	Tue	Wed	Thu	Fri	Sat	Sun

Foreman:

Contract No:

Hours Lost Due To Bad Weather

Visitors

Weather Conditions

AM	PM

Schedule

Completion Date:

Days Ahead of Schedute:

Days Behind Schedute:

Problems / Delays

Safely Issues:

Accidents/Incidents:

Summary Of Work Performed Today

Signature:

Name:

Equipment On Site	No Units	Working	
		Yes	No

Employee /Contractor	Trade	Contracted Hours	Overtime

Materials Delivered	From & Rate	Equipment Rented	No Units

Notes:

Date:		**Day:**	Mon	Tue	Wed	Thu	Fri	Sat	Sun
Foreman:									
Contract No:									

Hours Lost Due To Bad Weather

Visitors

Weather Conditions

AM	PM

Schedule

Completion Date:

Days Ahead of Schedute:

Days Behind Schedute:

Problems / Delays

Safely Issues:

Accidents/Incidents:

Summary Of Work Performed Today

Signature:

Name:

Equipment On Site	No Units	Working	
		Yes	No

Employee /Contractor	Trade	Contracted Hours	Overtime

Materials Delivered	From & Rate	Equipment Rented	No Units

Notes:

Date:		**Day:**	Mon	Tue	Wed	Thu	Fri	Sat	Sun
Foreman:									
Contract No:									

Hours Lost Due To Bad Weather	Visitors

Weather Conditions

AM	PM

Schedule	Problems / Delays
Completion Date:	
Days Ahead of Schedute:	
Days Behind Schedute:	

Safely Issues:	Accidents/Incidents:

Summary Of Work Performed Today

Signature:	Name:

Equipment On Site	No Units	Working	
		Yes	No

Employee /Contractor	Trade	Contracted Hours	Overtime

Materials Delivered	From & Rate	Equipment Rented	No Units

Notes:

Date:		**Day:**	Mon Tue Wed Thu Fri Sat Sun
Foreman:			
Contract No:			

Hours Lost Due To Bad Weather

Visitors

Weather Conditions

AM	PM

Schedule

Completion Date:	
Days Ahead of Schedute:	
Days Behind Schedute:	

Problems / Delays

Safely Issues:

Accidents/Incidents:

Summary Of Work Performed Today

Signature:	Name:

Equipment On Site	No Units	Working	
		Yes	No

Employee /Contractor	Trade	Contracted Hours	Overtime

Materials Delivered	From & Rate	Equipment Rented	No Units

Notes:

Date:	**Day:**	Mon	Tue	Wed	Thu	Fri	Sat	Sun

Foreman:

Contract No:

Hours Lost Due To Bad Weather

Visitors

Weather Conditions

AM	PM

Schedule

Completion Date:

Days Ahead of Schedute:

Days Behind Schedule:

Problems / Delays

Safely Issues:

Accidents/Incidents:

Summary Of Work Performed Today

Signature:

Name:

Equipment On Site	No Units	Working	
		Yes	No

Employee /Contractor	Trade	Contracted Hours	Overtime

Materials Delivered	From & Rate	Equipment Rented	No Units

Notes:

| Date: | | **Day:** | Mon | Tue | Wed | Thu | Fri | Sat | Sun |

Foreman:

Contract No:

Hours Lost Due To Bad Weather

Visitors

Weather Conditions

AM	PM

Schedule

Completion Date:

Days Ahead of Schedute:

Days Behind Schedute:

Problems / Delays

Safely Issues:

Accidents/Incidents:

Summary Of Work Performed Today

Signature:

Name:

Equipment On Site	No Units	Working	
		Yes	No

Employee /Contractor	Trade	Contracted Hours	Overtime

Materials Delivered	From & Rate	Equipment Rented	No Units

Notes:

Date:		**Day:**	Mon	Tue	Wed	Thu	Fri	Sat	Sun

Foreman:	
Contract No:	

Hours Lost Due To Bad Weather

Visitors

Weather Conditions

AM	PM

Schedule

Completion Date:	
Days Ahead of Schedute:	
Days Behind Schedute:	

Problems / Delays

Safely Issues:

Accidents/Incidents:

Summary Of Work Performed Today

Signature:	Name:

Equipment On Site	No Units	Working	
		Yes	No

Employee /Contractor	Trade	Contracted Hours	Overtime

Materials Delivered	From & Rate	Equipment Rented	No Units

Notes:

| Date: | | **Day:** | Mon | Tue | Wed | Thu | Fri | Sat | Sun |

| Foreman: | |

| Contract No: | |

Hours Lost Due To Bad Weather

Visitors

Weather Conditions

AM	PM

Schedule

Completion Date:	
Days Ahead of Schedute:	
Days Behind Schedute:	

Problems / Delays

Safely Issues:

Accidents/Incidents:

Summary Of Work Performed Today

Signature:	Name:

Equipment On Site	No Units	Working	
		Yes	No

Employee /Contractor	Trade	Contracted Hours	Overtime

Materials Delivered	From & Rate	Equipment Rented	No Units
			Yes No

Notes:

www.ingramcontent.com/pod-product-compliance
Lightning Source LLC
LaVergne TN
LVHW041328200726